W9-APS-228

simple machines

Springs

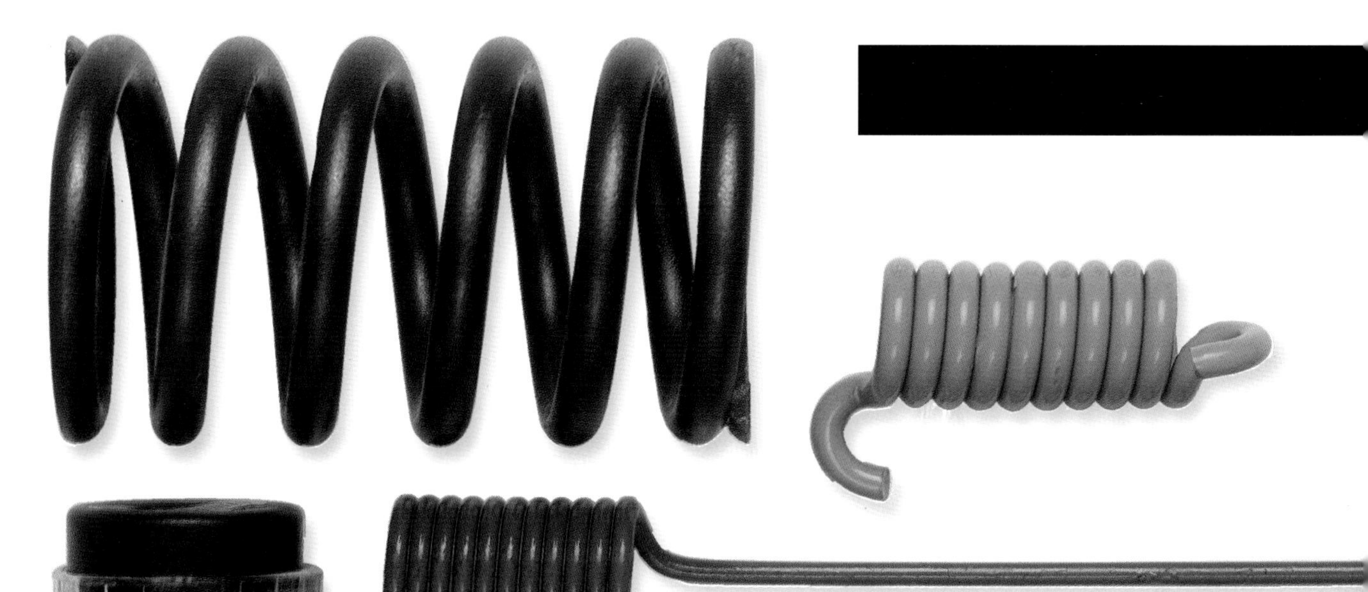

David Glover

This edition © 1997 Rigby Education
Published by Rigby Interactive Library,
an imprint of Rigby Education,
division of Reed Elsevier, Inc.
500 Coventry Lane
Crystal Lake, IL 60014

Printed in the United Kingdom

01 00 99 98 97
10 9 8 7 6 5 4 3 2 1

Library of Congress Cataloging-in-Publication Data
Glover, David.
Springs/David Glover.
p. cm. — (Simple machines)
Includes index.
Summary: Introduces the principles of springs as simple machines, using examples from everyday life.
ISBN 1-57572-082-5 (lib. bdg.)
1. Springs (Mechanism)—Juvenile literature. [1. Springs (Mechanism)] I. Title. II. Series.
TJ210.G55 1997
621.8 ' 11—dc20 96–15799
CIP
AC

Designed by Celia Floyd and Sharon Rudd
Illustrated by Barry Atkinson (pp. 9, 11, 19) and Tony Kenyon (pp. 7, 15, 23)

Acknowledgments
The publishers would like to thank the following for permission to reproduce photographs:
Trevor Clifford, pp. 4, 5, 6, 9, 12, 14, 15, 17, 20, 21, 22, 23; Sealy, p. 8; Colorsport, p.10; Spectrum Colour Library, p.16; Stockfile/Steven Behr, p.18; Zefa, p.19.

Cover photograph by Trevor Clifford

Note to the Reader

Some words in this book are printed in **bold** type. This indicates that the word is listed in the glossary on page 24. This glossary gives a brief explanation of words that may be new to you and tells you the page on which each word first appears.

Contents

What Are Springs? 4

Pogo Sticks. 6

Spring Beds and Chairs 8

Springboards 10

Door Springs and Locks 12

Spring Loaded 14

Spring Balances 16

Spring Wheels 18

Clockwork Springs 20

Pinballs and Cannonballs 22

Glossary 24

Index 24

Further Readings 24

What Are Springs?

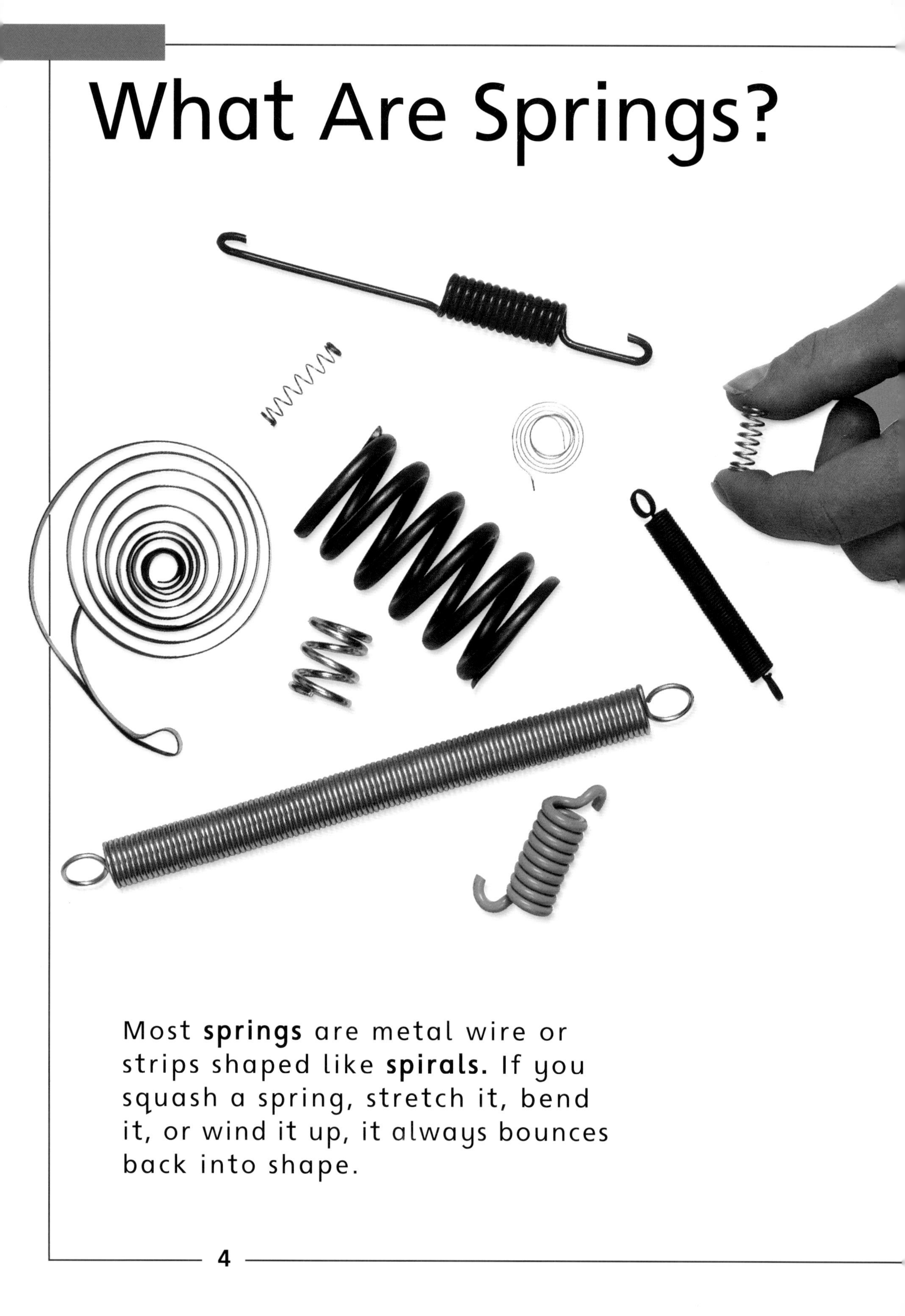

Most **springs** are metal wire or strips shaped like **spirals.** If you squash a spring, stretch it, bend it, or wind it up, it always bounces back into shape.

This jack-in-the-box has a spring inside it. When you push the clown into the box and fasten the lid, you are squashing and coiling its spring.

When you unfasten the lid, the spring uncoils, making the clown jump from its box. It can make you jump with surprise!

FACT FILE

Springs store the **energy** you use to squash them. When you let go of a squashed spring, it uses the stored energy to move things.

Pogo Sticks

A pogo stick is a pole with a strong spring and footrests on the bottom. As a person jumps, the spring bounces him or her up in the air.

Pogo crazy!

In 1985, Gary Stewart of Reading, Ohio, jumped 130,077 times on a pogo stick without falling off.

The spring on a pogo stick is a strong spiral of steel wire. Your weight squashes it down to make it shorter. The spring pushes back up as it tries to return to its normal shape. This force lifts you up into the air.

When you jump with a pogo stick, you squash the spring.

The spring pushes back up and bounces you into the air.

When you land, the spring squashes again, ready for the next bounce.

Spring Beds and Chairs

Have you ever bounced on your bed? Many bed mattresses are filled with wire springs. These help you to sleep comfortably at night. When you lie on the bed, the springs squash to support every part of your body.

The springs on this chair make the seat soft and comfortable. When you sit down, strong springs hold up your weight.

FACT FILE

Spring words

When a spring is made shorter, or coiled, we call the effect compression. When a spring is made longer, or stretched, we call the effect tension.

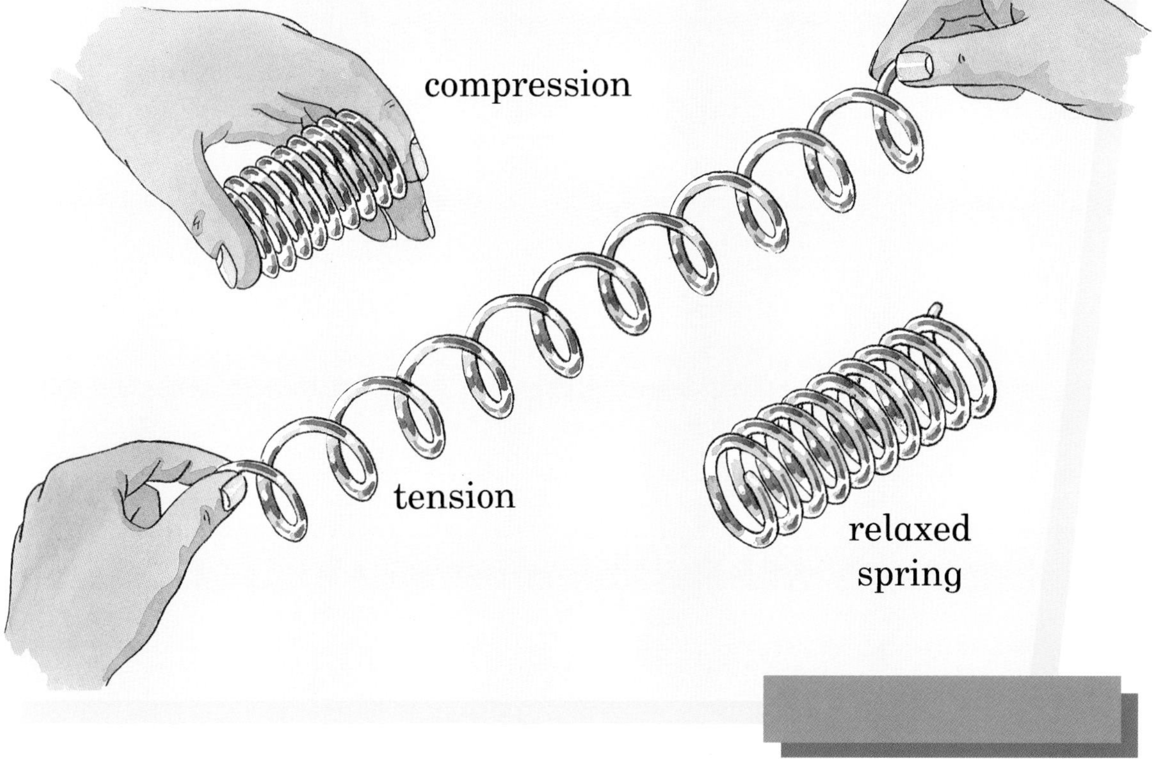

Springboards

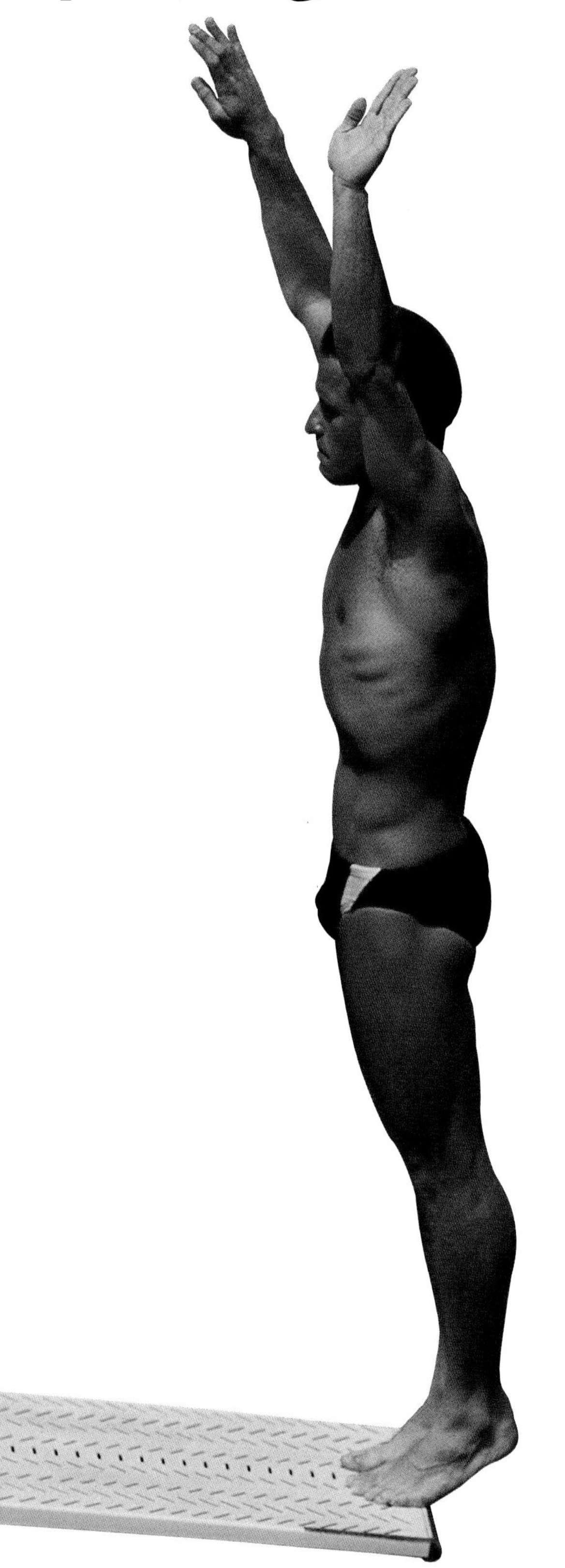

Springboards have the word *spring* in them, but they don't have springs. However, they are springy.

A diver bounces on a springboard at the pool. As he bounces, the board bends and then springs back to push him into the air.

A wheel on the side of the board moves back and forth when turned. This makes the springy part of the board shorter or longer to give different amounts of bounce.

This springboard is made from springy wood. The **gymnast** jumps on it firmly to help her spring high over the **vaulting horse**.

FACT FILE

A heavy person bends a springboard more than a light person. But when the board springs back, they would each go the same height. It takes more spring to lift a heavy person.

Door Springs and Locks

This gate has a spring at the top. The spring pulls the gate shut when someone leaves it open.

This door has a spring at the bottom. The spring stops the door from banging against the wall when someone pushes it too hard.

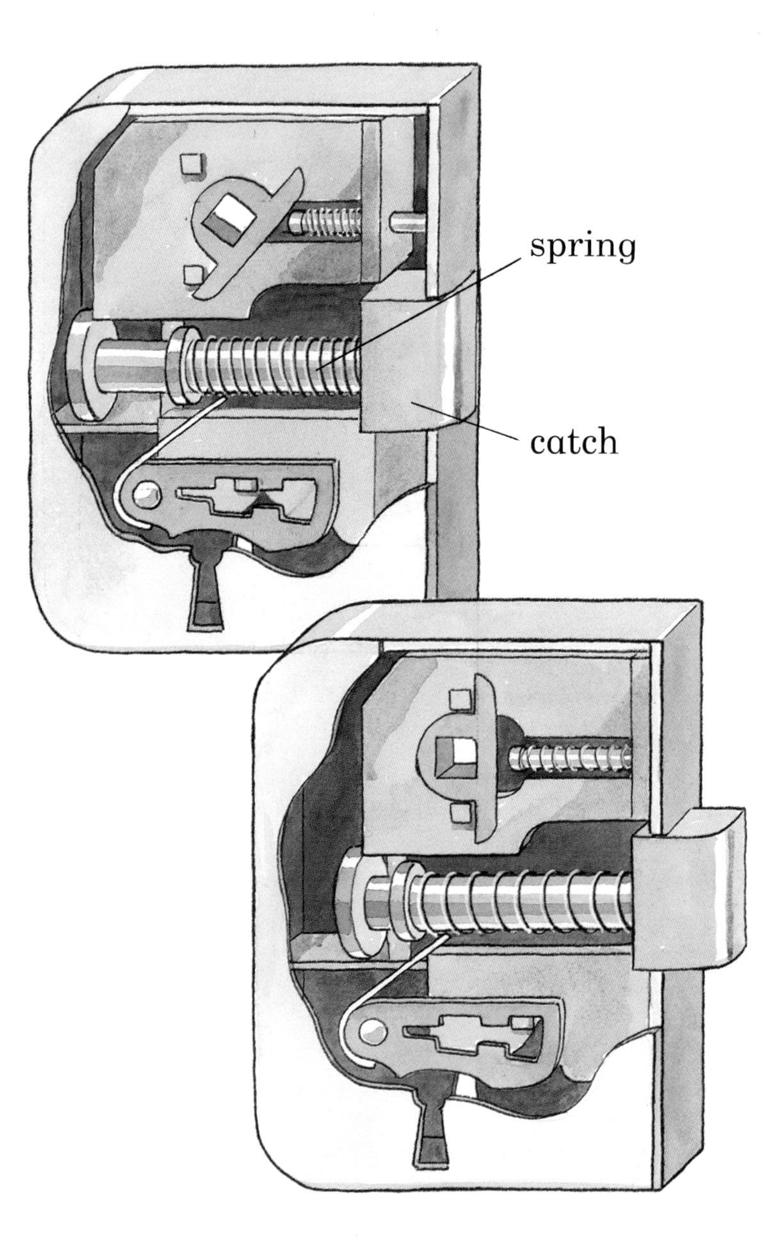

A door handle has a spring inside it. When you turn the handle, the spring pulls back the **catch**. When you let go, the spring pushes the catch back into place.

The door spring and the catch spring work together to keep the door closed.

FACT FILE

Old springs

The oldest existing springs were parts of old door locks. Some are more than 500 years old.

Spring Loaded

This pen is spring loaded. This means it has a spring inside it. When you press the button at the bottom of the pen, the tip you write with comes out of the case. A spring tries to push the tip back inside, but a catch holds it in place. When the button is pressed again, it releases the catch and the tip goes back inside the pen case.

Spring-loaded umbrellas are handy for carrying in a small bag. When it rains, you take off the cover, release the catch, and the umbrella springs open. A coiled spring inside the handle does all the work for you.

FACT FILE

Springy hats!

In the 1800s, some men wore spring-loaded top hats. When they went to the theater, they used to squash their tall hats and put them under the seats. When the play was over, they let the hats spring back into shape.

Spring Balances

A spring balance is a simple weighing machine that uses a spring. Spring balances are often used to weigh fish.

The weight of the fish stretches the spring. The stretched spring turns a needle around a **dial**. The heavier the fish, the more the spring stretches and the further the needle turns.

Some bathroom scales have a stiff spring inside. When you stand on one of these scales, the spring inside stretches a bit. This small movement is **exaggerated** by **levers**. The levers turn a dial to show your weight.

FACT FILE

The spring on a spring balance gets longer when you add more weight. A two-pound fish stretches the spring twice as much as a one-pound fish.

Spring Wheels

Some mountain bikes have springs in the forks that hold the front wheels. The springs squash and stretch when the wheels go over bumps. This makes the ride smoother and lets the rider go faster.

Moto-cross bikes speed around a very rough course. They leap high in the air over the hills and hit the ground again with a jolt. Long springs on the bike help to soften the landing for the rider.

Leaf springs

Special springs were invented for horse-drawn carts. They helped to make the passengers comfortable. The springs were made from thin strips of metal stacked together like the leaves of a book.

Clockwork Springs

Some old clocks must be wound with a key. The key winds up a spring inside the clock. As the wound-up spring slowly unwinds, it turns the hands of the clock.

The inside of a clock

Clockwork toys move because inside they have clock springs that turn a **motor.** A long time ago, many toys were powered by clockwork motors. Today, most moving toys have electric motors that are powered by **batteries.**

FACT FILE

A wind-up radio!

A clockwork radio that doesn't need batteries was recently invented. A clockwork motor turns a tiny machine called a *dynamo.* The dynamo makes electricity. You can wind up the radio when you want to listen.

Pinballs and Cannonballs

In a pinball machine, a spring shoots balls across a table. When you pull on the plunger, it coils the spring. When you let go, the spring uncoils and pushes the ball up the table. Players score when the ball hits targets on the table on its way back down.

The human cannon ball is a popular circus act. It looks as if the clown is fired from a gun by **gunpowder**, but the flash is caused by fireworks. It is really a big spring inside the cannon that pushes the clown up into the air.

FACT FILE

A staple gun has a strong spring inside. When the handle is squeezed, the spring is squashed. Then a catch releases the spring, and the gun fires a staple to fasten two things together.

Glossary

batteries Small packages of chemicals that make electricity 21

catch Piece of metal that clicks into a slot to keep a door closed 13

compression Result of an object being squeezed or squashed 9

dial Part of a scale where a pointer moves along a row of numbers to show the weight 16

energy Power 5

exaggerated Increased 17

gunpowder Powdered chemicals that burn with a flash and a bang 23

gymnast Acrobat 11

levers Bars which turn around a hinge or pivot 17

motor Machine that uses electricity or fuel to make things move 21

spiral Special shape that goes along and around at the same time. 4

spring Metal or plastic wires shaped like spirals that keep their shape even after being squashed or stretched 4

tension Result of an object being pulled or stretched 9

vaulting horse Special box that gymnasts jump and somersault over 11

Index

bikes 18–19

clock springs 20–21

compression 9

door locks 12–13

jack-in-the-box 5

leaf springs 19

mattress springs 8

pinball machines 22

pogo stick 6–7

spiral 4, 7

spring 4–23

spring balance 16–17

springboard 10–11

staple gun 23

tension 9

spring-loaded umbrella 15

spring-loaded pens 14

Further Readings

Barton, Byron. *Machines at Work.* New York: HarperCollins, 1987.

Stine, Megan. *Hands-On Science: Fun Machines.* Milwaukee: Gareth Stevens, 1993.